Poetry

plus

MaxD Physics

A New Physics with an Exact-Numbers Base

Has Physics Stressed You Out?
See How Poetry Breaks Might help.
Or. . . Might Physics Improve Your Poetry?

Edwin E. Hatch

i

ISBN 978-1-66784-221-9

Published in the United State of America
by Edwin E. Hatch

Introduction to "Just for a Break" Poetry

In the late 1980s my brother Ron, (Ronald R, Hatch, known worldwide for his contributions to GPS, probably the most universal implementation of precise scientific understanding ever devised.) told me of some of the knowledge of fundamental physics that was essential to the successful implementation of the precision of our modern GPS. We all know that precision is the most appreciated feature of GPS.

I have since spent over thirty years developing my own respect for precision, hoping to contribute to comparable precision in the numbers at the very base of modern physics. The best possible ending of that effort has been realized. I have moved from precise to exact—a precision I had only dreamed of.

But there was a price to be paid. Nearly thirty years of disciplined effort—from retirement in 1991 to realization of a goal in 2019, had been fun, but not easy. Sometimes I needed a break. I frequently took advantage of a large book of poetry—Robert Frost's "A Road not Taken", was a favorite of mine.

But reading was often not enough. Between 2001 and 2019 I played at using writing as a break and defined my supposed motivation in a few lines so typical of an amateur poet, as follows.

Let Words Play

July 2004:

(A rationale for leaving *Winter's Show and Tell* in my files.)

Not in past, nor yet in future,

Life is lived in present tense.

I "smell the roses", "seize the day",

Why can't I then, just let words play?

Of course, this attitude and the end results, could never justify publication or even retention in my files. But the writing did provide what I wanted at the time; a much-needed break. I have not included the major portion of what I had scribbled over the last twenty years. But if you can tolerate novice poetry, read on—just don't claim that I didn't warn you. *An early "break poem" follows.*

If I Were a Poet . . .

by Ed Hatch

If I were a poet, I think I might write:

Of beauty, of truth, of health and of wealth;

Of coming, of going, of time and its flowing;

Of giving, of getting, of sharing—or not;

Of life and of love, of yearning and learning;

Of beginnings and endings, of youth and old age;

Of houses, of homes, of far away places;

Of winter, of spring, of summer and fall;

Of morning, of evening, of day and of night;

Of flowers and trees, of birds and of bees;

Of paths that are steep, of paths that are rough;

Of peaceful green valleys, of cold stormy mountains;

Of hope for the future, of pain not yet past;

Of things soon discarded, of memories that last;

Of strength and of weakness, of health and desire;

Of laughing and crying, of living and dying;

Of what is and what's not, what's free and what's bought;

Of thinking, of knowing, of sharing, of showing;

Of praise and of promise, of being and growing;

Of winning, of losing, of options, decisions;

Of "Why?" and "Why not?", of risk and reward;

Of laughing, of crying, of joy and of sorrow;

Of light and of shadow, bright day and dark night;

Of things that are real and things that are not;

Of value, of worth, of status, of cost;

Of family, of friends, of strangers and neighbors;

Of fathers and mothers, of sons and of daughters;

Of brothers and sisters, of in-laws and others;

Of families made stronger, of families destroyed;

Of arriving, of meeting, of staying, of leaving;

Of dancing, of singing, of passionate living;

Of the rich, of the mighty, of greatest and least;

Of strength and of struggles (whether lost or hard won);

Of surplus, of shortage, of forethought, of haste;

Of engaged or estranged, of spellbound or bored;

And last, but not least, of the pen and the sword.

If I were a poet . . .   Aren't you glad that I'm not?

While enjoying physics as I never dreamed I could, other breaks were needed.

Winter's Show and Tell
by Ed Hatch

Cascading Shtick Format

A sheer delight
  in wintry night
    is never trite.
      (Yes, there's a bite
        though very slight.)

No snow in sight,
  yet shades of white
    leave scene so bright
      that shadows fight
        with all their might
          to claim the right
            to quench the light.

As if to please
  the eye that sees
    the night-wind frees
      a gentle breeze
        and air-stream seas
          with playful ease
            begin to tease
              the tallest trees.

While others stay
  like limbs of clay
    high branches sway
      (to, then away)
        as if to say:
          "C'mon let's play.
            You can, you may,
              you know the way.
                Why wait for day?"

Now Winter's grasp
  brings icy blast,
    leaves all aghast,
      as moonbeams cast
        dark shadows fast
          across the grass,
            black clouds rush past,
              light will not last.

Can wintry will
  the darkness fill
    with ghostly chill?
      The breezes kill?
        Stark night goes still,
          but just until
            clouds crest the hill.

And now, although
  the night-winds grow
    to dreadful blow
      and trees bend low
        'neath massive flow
        of driving snow,
          a soothing glow
            persists below
              the wintry throw.
We may be slow,
  but even so,
    we surely know
      that "tell" must go;
        it's time for show.
See Autumn fold,
  give up Her gold,
    as storm yet bold,
      though hours old,
        brings Winter's hold
        of crippling cold.
          For days untold,
            the winds yet scold.
            Her show behold,
            Her tell is told.

The Critics are not kind:

J. T. Kite
A "wintry night"
  and "shadow's fight
    with all their might"?
      The tone's not right.

Francis Frieze
"the night wind frees
  a gentle breeze"
    and "air-stream seas"?
      Oh please!
        This reader sees,
          and promptly flees.

Susan Pabst
"dark shadows fast
across the grass"?
Pathetic grasp:
This work won't last.

William (Bill) Hill
"Can wintry will
the darkness fill
with ghostly chill?
The breezes kill?"
What lack of skill!

Adam Grey
What can one say
to "limbs of clay"?
Some readers may
accept and stay,
but I 'm away.

Autumn Seybold
"Her tell is told"?
I find no gold.
As scenes unfold
the form gets old
and leaves one cold.

The following Picture-Poem—my first ever, was an enjoyable series of every-day breaks.

There are a few things very different about this coming last poem. Our black Labrador dog died of cancer in 2012. After a few days of grieving, we decided we still needed a dog. At a local rescue center, we found Tally, the name short for Tallulah Bankhead, an actress famous for the striking beauty of her eyes and eyebrows. That was Tally. So we gave this recue dog a forever home.

Shortly after bringing her home, I was taking her to a local dog park every morning at dawn. She made a lot of friends, both other dogs and other people. She also began climbing one of the trees every day. That was rather fun until with the approach of fall squirrels were usually in the park. Tally loved them and if one was also in her tree, she refused to come down to go home. With my physics effort at its peak this was irritating.

But she loved other dogs and other people. Friends took lots of pictures of her in what became her tree, and I used some to create the following picture-poem, with her as the star. It is not great, but making it was responsible for some of my biggest breaks from my physics. *It was taking too much time.*

Enjoy if you can.

Tally's Tree
by Ed Hatch

Tally's tree is in the

Auburn Ravine Dog Park – Lincoln, California

Squirrels climb trees, dogs do not.

Before we met Tally, that's what we all thought.

There's something up there, up there in that tree!

What can that be looking straight down at me?

A dog?   No!   No!   Dogs never climb trees!

I cannot believe what I think that I see!

But then when it moved, it was easy to see,

It was a dog! A dog in a tree!

Easy to see—yes, easy to see!

But how could it be—a dog in a tree?

And then the dog laughed: Why stare at me?

I'm just a dog—just a dog in my tree.

I was scared—did that dog talk to me?

Then she just laughed: "It's really just me—just a dog in a tree."

Don't you like dogs? Come up and play.

If you need help, I can show you the way.

What a weird dog. How strange she must be!

Why in the world is she up in that tree?

I love to watch you – watch you watching me.

Wouldn't you love to be here in my tree?

I see things from up here I can't see from down there.

Squirrels and birds; other dogs everywhere.

See that squirrel up there? What does he see?

He knows I'm a dog and he's laughing at me!

There are birds up there in the top of my tree.

I can see them from here. Come watch them with me.

The birds and the bees—and squirrels too!

What a wonderful world! Such a great view!

I see people and dogs — friends that I love!

So much to see, when seen from above!

This is my world! I know that is true!

Come see for yourself, it's your world too!

My tree's in a byway, the Pacific Coast Flyway,

From here I watch geese as they travel their Skyway

From Tally with Love

Keep looking up when you visit my park.
You might see a lot! I once saw a lark!

It was quite a sacrifice, but Tally really wanted to make room for her favorite poem. So here it is on your right—one last poem.

She wants to tell you that she wishes she could have been with you as you viewed her pictures and her responses to the people watching her.

In any case, this is the end of the "just take a break" poetry. If you would like to be introduced to the exciting new world of theoretical physics, just proceed through the following pages.

With Tally, I hope you have enjoyed what you have seen and, if interested, enjoy the physics also.

Wishing you and your family the very best,

Ed Hatch

Tally requested one last poem.

Please think of her as you read this poem. I really don't know why she loved it so. Maybe because a dog's life is much shorter than ours? I really have no idea. Here it is.

As Evening Comes

by Ed Hatch

There are some things I'd like to write:
But though I know both sound and sight,
It's hard to find a way to say
What I have learned from day to day.
Yes, there's a song I'd like to share,
Free verse or rhyme—I know it's there.

But where is there? I feel a tide,
An ebb and flow—it's quite a ride.
One moment up, one moment down,
A day of joy—and time moves on.
Some things, I've learned, I cannot know.
To see, to hear—I pause, then go.

But where to go? Paths soon diverge,
Each leading on to who knows where.
Paths "not taken", choices made,
Dreams there are, but dreams may fade.
As tide goes out, I must move on,
Some battles lost, some prizes won.

Changing tides brings stress and strain,
And as years pass regrets remain.
Yet past is past. When day is done,
I only hope I've helped someone
Embrace the tide—enjoy each wave;
To love to live! . . . too soon the grave.

An introduction to MaxD; the Very Core of Modern Physics

Copyright © 2022, Edwin E. Hatch

Establishing the MaxD Core

In 1916 Arnold Sommerfeld derived the fine structure constant in electrostatic cgs units as: $(e^2 / (\hbar c))$.

- We now know that $(\hbar c) = ((m_e c^2 r_e) / \alpha)$, so that $(e^2 / (\hbar c))$ clearly reduces to: $(\alpha e^2) / (m_e c^2 r_e)$.

Where $e^2 = (m_e c^2 r_e)$, so that, as Sommerfeld proposed, α does indeed equal $(e^2 / (\hbar c))$. We offer proof:

- In cgs units: $E_G = e^2$ $\qquad\qquad = (m_e c^2 r_e) = m_e (r_e^3 / t_e^2) = 2.3070775523377E\text{-}19$.
- In SI units: $E_G = e^2 / (4pi\ \varepsilon_0)$ $= e^2 (c^2 \times 10^{-7}) = (m_e c^2 r_e) = m_e (r_e^3 / t_e^2) = 2.3070775523377E\text{-}28$.

E_G gives us the EN-Core, the Exact-Numbers Core of all of physics.

The Electro-Gravito, E_G, label is intended to highlight its general applicability to the MaxD Core.

- MaxD is based on Maxwell's Dimensions of mass, $[L^3 T^{-2}]$, and of gravity's G, $[L^3 T^{-2} M^{-1}]$.
- E_G is the MaxD-Ratio Numerator. This Ratio is the key to the definitions of EN-Core relationships.
- The fine structure constant, α, is exactly $((v_B / c) \times (r_e / r_B) \times (t_e / t_B))^{(1/6)}$: $\alpha = 0.00729735256929315$.

MR, The MaxD Ratios		The EN-Core Values		The Constants
$(m_e c^2 r_e)$ / (r_e^3 / t_e^2)		= 9.1093837015E-31	= m_e	The rest mass of the electron
$(m_e c^2 r_e)$ / $(m_e c\ r_e)$		= 299792458	= c	The speed of light
$(m_e c^2 r_e)$ / $(m_e c^2)$		= 2.8179403262E-15	= r_e	The classical radius of the electron
$(m_e c^2 r_e)$ / $(m_e c^3)$		= 9.39963715231288E-24	= t_e	The time t_e (length / speed), (r_e / c)
Valid MR numerator definitions are compared below. **The Numerator values are always identical to E_G.**				
$(m_e c^2 r_e)$ / $(m_e v_B r_B)$		= 2187691.26364101	= v_B	The Bohr velocity
$(m_e c^2 r_e)$ / $(m_e v_B^2)$		= 5.29177210902967E-11	= r_B	The Bohr radius
$(m_e c^2 r_e)$ / $(m_e v_B^3)$		= 2.41888432658569E-17	= t_B	The Bohr time
$(m_e c^2 r_e)$ / v_B		= 1.05457181764213E-34	= $\hbar$	Planck's reduced constant
$(m_e c^2 r_e)$ / $(\hbar c)$		= 0.00729735256929315	= α	The fine structure constant
$(m_e c^2 r_e)$ / $(c^2 \times 10^{-7})$		= 2.56696996792859E-38	= e^2	The square of the electron's charge
$(m_e c^2 r_e)$ / r_B		= 4.3597447221905E-18	= E_h	The Hartree energy
The MaxD Ratio Numerators, MRNs, are always MRN = E_G—The denominator determines the MR value.				

As an aside, I love the way MaxD-Ratio Physics presents the relationship between Planck's $\hbar$ and my MRNs.

- Planck's $\hbar$ is: $(m_e c^2 r_e) / v_B$ $= (m_e c^2) \times (r_e / v_B)$. $\qquad$ This is (*Energy x Time*).
- The MRN is then: $(m_e c^2 r_e) / v_B)\ v_B = (m_e c^2) \times (r_e / v_B) \times v_B$. $\qquad$ This is (*Energy x Time x Velocity*).

Note that the constant factors of our EN-Core are few and well defined:

- m_e is the mass of the electron. $\qquad\qquad$ c is the speed of light, $c = (r_e / t_e)$.
- r_e is the classical radius of the electron. $r_e = (c\ t_e)$. $\quad$ t_e is the time light takes to traverse r_e, $t_e = (r_e / c)$.

By expressing Maxwell's dimensions of Newton's G in terms of electron units we have:

- The Hatch Gauge constant, $G_H = (r_e^3 / t_e^2) / m_e$. $\qquad$ This is Maxwell's dimensions of G.
- $(G_H m_e) = (r_e^3 / t_e^2)$ $\quad = (c^2 r_e)$ $= (c^3 t_e)$. $\qquad$ This is Maxwell's dimensions of mass.

Finally, "We must accept Nature as she is". The physics community is now being tested on this point.

G_H, the Hatch Gauge constant, is used to derive the constants of the Bohr model of the hydrogen atom. Among many other valuable insights, the Bohr model presents G_H as a conversion constant.

- G_H first derives m_e as *the dimensions of the mass m_e:* $(G_H\ m_e) = (r_e\ c^2) = (t_e\ c^3) = (r_e^3 / t_e^2)$
- Then *replaces these dimensions of mass* with m_e *itself:* $(r_e^3 / t_e^2) / G_H$ gives m_e

The derivation details of the ground-state Bohr Model, with $(v = v_B)$, follows.

	Derivation	Simplifications		Net Dimensions		Label
r_B	$= (G_H\ m_e) / v_B^2$	$= (r_e\ c^2) / v_B^2$	$= r_e / \alpha^2$	r_e	$[L]$	The Bohr radius
t_B	$= (G_H\ m_e) / v_B^3$	$= (t_e\ c^3) / v_B^3$	$= t_e / \alpha^3$	t_e	$[T]$	The Bohr time
v_B	$= (r_B / t_B)$	$= c / (c / v_B) = v_B$	$= (c\ \alpha)$	c	$[L\ T^{-1}]$	The Bohr velocity
m_B	$= (r_B^3 / t_B^2) / G_H$	$= m_e$	$= m_e$	m_e	$[L^3\ T^{-2}]$	Electron's mass
C_B	$= (r_B^4 / t_B^2) / G_H$	$= (m_e\ c^2\ r_e) / v_B^2$	$= (m_e\ r_e) / \alpha^2$	$(m_e\ r_e)$	$[L^4\ T^{-2}]$	Coulomb constant
$\hbar$	$= (r_B^5 / t_B^3) / G_H$	$= (m_e\ c^2\ r_e) / v_B$	$= (m_e\ c\ r_e) / \alpha$	$(m_e\ c\ r_e)$	$[L^5\ T^{-3}]$	Plank's $\hbar$
F_B	$= (r_B^4 / t_B^4) / G_H$	$= (m_e\ v_B^2) / r_B$	$= (m_e\ r_e / t_e^2)\ \alpha^4$	$(m_e\ r_e / t_e^2)$	$[L^4\ T^{-4}]$	The Bohr force
E_h	$= (r_B^5 / t_B^4) / G_H$	$= (m_e\ v_B^2)$	$= (m_e\ c^2)\ \alpha^2$	$(m_e\ v_B^2)$	$[L^5\ T^{-4}]$	Hartree energy

The Bohr model gives values, dimensions, and the nature of the fine structure constant, α.

Extending the Picture. With $v = (v_B / n)$, the integer n gives us the nth energy-level values.

Ground-state, $v = v_B$		Any n	
$G_H\ m_e\ / v_B^2$	$= r_B$	$r_{Bn} = G_H\ m_e\ / (v_B / n)^2$	$= (r_B\ n^2)$
$G_H\ m_e\ / v_B^3$	$= t_B$	$t_{Bn} = G_H\ m_e\ / (v_B / n)^3$	$= (t_B\ n^3)$
(r_B / t_B)	$= v_B$	(r_{Bn} / t_{Bn})	$= (v_B / n)$
$(r_B^3 / t_B^2) / G_H$	$= m_e$	$(r_{Bn}^3 / t_{Bn}^2) / G_H$	$= m_e$
$(r_B^4 / t_B^2) / G_H$	$= C_B$	$(r_{Bn}^4 / t_{Bn}^2) / G_H$	$= (C_B\ n^2)$
$(r_B^5 / t_B^3) / G_H$	$= \hbar$	$(r_{Bn}^5 / t_{Bn}^3) / G_H$	$= (\hbar\ n)$
$(r_B^4 / t_B^4) / G_H$	$= F_B$	$(r_{Bn}^4 / t_{Bn}^4) / G_H$	$= (F_B / n^4)$
$(r_B^5 / t_B^4) / G_H$	$= E_h$	$(r_{Bn}^5 / t_{Bn}^4) / G_H$	$= (E_h / n^2)$

The Bohr model tells an interesting story.

G_H derivations of binding energies at every level use: $v = (v_B / n)$ $r = (r_B\ n^2)$ $t = (t_B\ n^3)$

Binding energy in electron-volts is given by $((r^5 / t^4) / G_H) / (2\ e)$

n	Derivation	Derived	Measured
1	$((r^5 / t^4) / G_H) / (2\ e)$	$= 13.6056931192505$	13.60 *eV*
2	$((r^5 / t^4) / G_H) / (2\ e)$	$= 3.40142327981263$	3.40 *eV*
3	$((r^5 / t^4) / G_H) / (2\ e)$	$= 1.51174367991672$	1.511 *eV*
4	$((r^5 / t^4) / G_H) / (2\ e)$	$= 0.850355819953157$	0.850 *eV*
5	$((r^5 / t^4) / G_H) / (2\ e)$	$= 0.54422772477002$	0.544 *eV*
6	$((r^5 / t^4) / G_H) / (2\ e)$	$= 0.377935919979181$	0.378 *eV*
7	$((r^5 / t^4) / G_H) / (2\ e)$	$= 0.277667206515316$	0.278 *eV*

Note the agreement with measured values. Consistent and exact numbers provide the MaxD-Ratio base.

We now have the foundation and much of the structure of MaxD Physics.

Our Universe: Exact.   Consistent.   Definitive.   Beautiful.

Once more. The Bohr model of the hydrogen atom uses G_H as a conversion constant, where:

- G_H first derives m_e as *the dimensions of m_e*: $(G_H\, m_e) = (r_e\, c^2) = (t_e\, c^3) = (r_e^3 / t_e^2)$
- Then *replaces the dimensions of m_e with m_e itself*: $(r_e^3 / t_e^2) / G_H$ gives m_e

Exact values of the ground-state G_H derivations of the Bohr model are as follows.

Label and Derivation		The Derived Ground-State Constant	Value
r_B	$= r_e / \alpha^2$	The Bohr radius—Atomic unit of length	= 5.29177210902969E-11
t_B	$= t_e / \alpha^3$	The Bohr time—Atomic unit of time	= 2.4188843265857E-17
v_B	$= (r_B / t_B)$	The Bohr Velocity—Atomic unit of velocity	= 2187691.26364101
m_B	$= (r_B^3 / t_B^2) / G_H$	Rest mass of the electron	= 9.1093837015E-31
C_B	$= (r_B^4 / t_B^2) / G_H$	A Coulomb constant	= 4.82047826020473E-41
$\hbar$	$= (r_B^5 / t_B^3) / G_H$	Planck's reduced constant	= 1.05457181764213E-34
F_B	$= (r_B^4 / t_B^4) / G_H$	The Bohr force – Atomic unit of force	= 8.23872349822318E-8
E_h	$= (r_B^5 / t_B^4) / G_H$	The Hartree energy	= 4.35974472219049E-18

Below, CODATA values are compared to the preferred exact MaxD-Ratio values.

CODATA and MaxD Ratio		Contrasted Values	The Constant
2018 CODATA		= 2187691.26364	The Bohr velocity, Atomic Unit of Velocity
$v_B\,(m_e\, v_B\, r_B)$	$/ (m_e\, v_B\, r_B)$	= 2187691.26364101	
2018 CODATA		= 5.29177210903E-11	The Bohr radius, Atomic Unit of Length
$r_B\,(m_e\, v_B^2)$	$/ (m_e\, v_B^2)$	= 5.29177210902967E-11	
2018 CODATA		= 2.4188843265857E-17	The Bohr time, Atomic Unit of Time
$t_B\,(m_e\, v_B^3)$	$/ (m_e\, v_B^3)$	= 2.41888432658569E-17	
2018 CODATA		= 1.054571817E-34	Planck's reduced constant
$\hbar\,(c\, a)$	$/ (c\, \alpha)$	= 1.05457181764213E-34	
2018 CODATA		= 0.0072973525693	The fine structure constant
$\alpha\,(\hbar\, c)$	$/ (\hbar\, c)$	= 0.00729735256929315	
2018 CODATA		= 2.56696996653557E-38	The square of the electron's charge
$e^2\,(c^2 \times 10^{-7})$	$/ (c^2 \times 10^{-7})$	= 2.56696996792859E-38	
2018 CODATA		= 4.3597447222071E-18	The Hartree energy
$E_h\, r_B$	$/ r_B$	= 4.3597447221905E-18	

Again, the denominator of the MaxD ratio determines the value given.

Some examples of how the MaxD-Ratio derivations are defined.

	The MaxD Ratio Numerator, MRN			The MR denominator completes the picture		
m_e	$m_e (r_e c^2)$	= 2.3070775523377E-28	$m_e (r_e c^2)$	/ $(r_e c^2)$	= 9.1093837015E-31
r_e	$r_e (m_e c^2)$	= 2.3070775523377E-28	$r_e (m_e c^2)$	/ $(m_e c^2)$	= 2.8179403262E-15
t_e	$t_e (m_e c^3)$	= 2.3070775523377E-28	$t_e (m_e c^3)$	/ $(m_e c^3)$	= 9.39963715231288E-24
r_B	$r_B (m_e v_B^2)$	= 2.3070775523377E-28	$r_B (m_e v_B^2)$	/ $(m_e v_B^2)$	= 5.29177210902967E-11
t_B	$t_B (m_e v_B^3)$	= 2.3070775523377E-28	$t_B (m_e v_B^3)$	/ $(m_e v_B^3)$	= 2.41888432658569E-17
v_B	$v_B (m_e r_B v_B)$	= 2.3070775523377E-28	$v_B (m_e r_B v_B)$	/ $(m_e r_B v_B)$	= 2187691.26364101
$\hbar$	$\hbar v_B$	= 2.3070775523377E-28	$\hbar v_B$	/ v_B	= 1.05457181764213E-34
α	$\alpha (\hbar c)$	= 2.3070775523377E-28	$\alpha (\hbar c)$	/ $(\hbar c)$	= 0.00729735256929315
e^2	$e^2 (c^2 \times 10^{-7})$	= 2.3070775523377E-28	$e^2 (c^2 \times 10^{-7})$	/ $(c^2 \times 10^{-7})$	= 2.56696996792859E-38
E_h	$E_h r_B$	= 2.3070775523377E-28	$E_h r_B$	/ r_B	= 4.3597447221905E-18
G	$G (m_e^2 S_G)$	= 2.3070775523377E-28	$G (m_e^2 S_G)$	/ $(m_e^2 S_G)$	= 6.67430E-11
G_H	$G_H (m_e^2)$	= 2.3070775523377E-28	$G_H (m_e^2)$	/ (m_e^2)	= 2.7802522591364E+32
S_G	$S_G (m_e^2 G)$	= 2.3070775523377E-28	$S_G (m_e^2 G)$	/ $(m_e^2 G)$	= 4.16560876666677E+42
G_P	$G_P (m_e^2 / S_G)$	= 2.3070775523377E-28	$G_P (m_e^2 / S_G)$	/ (m_e^2 / S_G)	= 1.15814431842037E+75
F_N	$F_N (r_e^2 S_G)$	= 2.3070775523377E-28	$F_N (r_e^2 S_G)$	/ $(r_e^2 S_G)$	= 6.97461325379537E-42
F_H	$F_H r_e^2$	= 2.3070775523377E-28	$F_H r_e^2$	/ r_e^2	= 29.0535101141202
F_P	$F_P (r_e^2 / S_G)$	= 2.3070775523377E-28	$F_P (r_e^2 / S_G)$	/ (r_e^2 / S_G)	= 1.21025556433821E+44

***

Once more the ground-state derivations of the Bohr model tell a surprising story.

r_B	= $G_H m_e / v_B^2$	= $r_e (c^2 / v_B^2)$	= r_e / α^2	Bohr radius	The Atomic Unit of Length
t_B	= $G_H m_e / v_B^3$	= $t_e (c^3 / v_B^3)$	= t_e / α^3	Bohr time	The Atomic Unit of Time
v_B	= (r_B / t_B)	= $c / (c / v_B) = v_B$	= $(c \alpha)$	Bohr velocity	The Atomic Unit of Velocity
These derivations show conclusively that, (c / v_B) is $(1 / \alpha)$, and therefore, $\alpha = (v_B / c)$: But . . .					
Fact: $(r_B^3 / t_B^2) / G_H$	$r_B^3 = (r_e^3 / \alpha^6)$	$t_B^2 = (t_e^2 / \alpha^6)$	$(r_B^3 / t_B^2) = (r_e^3 / t_e^2)$	$(r_e^3 / t_e^2) / G_H = m_e$	
$(r_B^3 / t_B^2) / G_H = (r_e^3 / t_e^2) / G_H = m_e$: dividing by G_H simply replaces the (r_e^3 / t_e^2) with m_e.					

Fact: $(r_B^5 / t_B^3) / G_H = \hbar$	m_e is first extracted leaving	(r_B^2 / t_B)	= $(c r_e / \alpha)$	$\hbar = (m_e c r_e) / \alpha$

Fact: $(r_B^5 / t_B^4) / G_H = E_h$	m_e is first extracted leaving	(r_B^2 / t_B^2)	= $(c^2 \alpha^2)$	$E_h = (m_e v_B^2)$

The three "Gauge Constants" and the associated "Force Constants" provide a surprising picture.

Newton Gauge, $G = (G_H / S_G)$		Hatch Gauge, $G_H = (r_{e3} / t_{e2}) / m_e$	Planck Gauge, $G_P = (G_H S_G)$	
F_N	= (F_H / S_G)	= The Newton Force The Planck Force =	= c^4 / G)	= F_P
F_H	= $G_H (m_e^2 / r_e^2)$	= The Hatch Force =	= c^4 / G_H)	= F_H
F_P	= $(F_H S_G)$	= The Planck Force The Newton Force =	= c^4 / G_P	= F_N

We now look closer at the MaxD-Ratio and denominator specified relationships.

The following tables demonstrate that relationships are defined by the MaxD Ratio Denominator.

	Planck's Reduced Constant, $\hbar$		The Fine Structure Constant, α	
m_e	$m_e\,(r_e\,c^2)\,/\,v_B$	$= 1.05457181764213\text{E-}34$	$m_e\,(r_e\,c^2)\,/\,(\hbar\,c)$	$= 0.00729735256929315$
r_e	$r_e\,(m_e\,c^2)\,/\,v_B$	$= 1.05457181764213\text{E-}34$	$r_e\,(m_e\,c^2)\,/\,(\hbar\,c)$	$= 0.00729735256929315$
t_e	$t_e\,(m_e\,c^3)\,/\,v_B$	$= 1.05457181764213\text{E-}34$	$t_e\,(m_e\,c^3)\,/\,(\hbar\,c)$	$= 0.00729735256929315$
r_B	$r_B\,(m_e\,v_B^2)\,/\,v_B$	$= 1.05457181764213\text{E-}34$	$r_B\,(m_e\,v_B^2)\,/\,(\hbar\,c)$	$= 0.00729735256929315$
t_B	$t_B\,(m_e\,v_B^3)\,/\,v_B$	$= 1.05457181764213\text{E-}34$	$t_B\,(m_e\,v_B^3)\,/\,(\hbar\,c)$	$= 0.00729735256929315$
v_B	$v_B\,(m_e\,r_B\,v_B)\,/\,v_B$	$= 1.05457181764213\text{E-}34$	$v_B\,(m_e\,r_B\,v_B)\,/\,(\hbar\,c)$	$= 0.00729735256929315$
$\hbar$	$\hbar\,v_B\,/\,v_B$	$= 1.05457181764213\text{E-}34$	$\hbar\,v_B\,/\,(\hbar\,c)$	$= 0.00729735256929315$
α	$\alpha\,(\hbar\,c)\,/\,v_B$	$= 1.05457181764213\text{E-}34$	$\alpha\,(\hbar\,c)\,/\,(\hbar\,c)$	$= 0.00729735256929315$
e^2	$e^2\,(c^2 \times 10^{-7})\,/\,v_B$	$= 1.05457181764213\text{E-}34$	$e^2\,(c^2 \times 10^{-7})\,/\,(\hbar\,c)$	$= 0.00729735256929315$
E_h	$E_h\,r_B\,/\,v_B$	$= 1.05457181764213\text{E-}34$	$E_h\,r_B\,/\,(\hbar\,c)$	$= 0.00729735256929315$
G	$G\,(m_e^2\,S_G)\,/\,v_B$	$= 1.05457181764213\text{E-}34$	$G\,(m_e^2\,S_G)\,/\,(\hbar\,c)$	$= 0.00729735256929315$
G_H	$G_H\,(m_e^2)\,/\,v_B$	$= 1.05457181764213\text{E-}34$	$G_H\,(m_e^2)\,/\,(\hbar\,c)$	$= 0.00729735256929315$
S_G	$S_G\,(m_e^2\,G)\,/\,v_B$	$= 1.05457181764213\text{E-}34$	$S_G\,(m_e^2\,G)\,/\,(\hbar\,c)$	$= 0.00729735256929315$
G_P	$G_P\,(m_e^2\,/\,S_G)\,/\,v_B$	$= 1.05457181764213\text{E-}34$	$G_P\,(m_e^2\,/\,S_G)\,/\,(\hbar\,c)$	$= 0.00729735256929315$
F_N	$F_N\,(r_e^2\,S_G)\,/\,v_B$	$= 1.05457181764213\text{E-}34$	$F_N\,(r_e^2\,S_G)\,/\,(\hbar\,c)$	$= 0.00729735256929315$
F_H	$F_H\,r_e^2\,/\,v_B$	$= 1.05457181764213\text{E-}34$	$F_H\,r_e^2\,/\,(\hbar\,c)$	$= 0.00729735256929315$
F_P	$F_P\,(r_e^2\,/\,S_G)\,/\,v_B$	$= 1.05457181764213\text{E-}34$	$F_P\,(r_e^2\,/\,S_G)\,/\,(\hbar\,c)$	$= 0.00729735256929315$

***

	The mass of the electron, m_e		The Symmetry Constant, S_G	
m_e	$m_e\,(r_e\,c^2)\,/\,(r_e\,c^2)$	$= 9.1093837015\text{E-}31$	$m_e\,(r_e\,c^2)\,/\,(m_e^2\,G)$	$= 4.16560876666677\text{E+}42$
r_e	$r_e\,(m_e\,c^2)\,/\,(r_e\,c^2)$	$= 9.1093837015\text{E-}31$	$r_e\,(m_e\,c^2)\,/\,(m_e^2\,G)$	$= 4.16560876666677\text{E+}42$
t_e	$t_e\,(m_e\,c^3)\,/\,(r_e\,c^2)$	$= 9.1093837015\text{E-}31$	$t_e\,(m_e\,c^3)\,/\,(m_e^2\,G)$	$= 4.16560876666677\text{E+}42$
r_B	$r_B\,(m_e\,v_B^2)\,/\,(r_e\,c^2)$	$= 9.1093837015\text{E-}31$	$r_B\,(m_e\,v_B^2)\,/\,(m_e^2\,G)$	$= 4.16560876666677\text{E+}42$
t_B	$t_B\,(m_e\,v_B^3)\,/\,(r_e\,c^2)$	$= 9.1093837015\text{E-}31$	$t_B\,(m_e\,v_B^3)\,/\,(m_e^2\,G)$	$= 4.16560876666677\text{E+}42$
v_B	$v_B\,(m_e\,r_B\,v_B)\,/\,(r_e\,c^2)$	$= 9.1093837015\text{E-}31$	$v_B\,(m_e\,r_B\,v_B)\,/\,(m_e^2\,G)$	$= 4.16560876666677\text{E+}42$
$\hbar$	$\hbar\,v_B\,/\,(r_e\,c^2)$	$= 9.1093837015\text{E-}31$	$\hbar\,v_B\,/\,(m_e^2\,G)$	$= 4.16560876666677\text{E+}42$
α	$\alpha\,(\hbar\,c)\,/\,(r_e\,c^2)$	$= 9.1093837015\text{E-}31$	$\alpha\,(\hbar\,c)\,/\,(m_e^2\,G)$	$= 4.16560876666677\text{E+}42$
e^2	$e^2\,(c^2 \times 10^{-7})\,/\,(r_e\,c^2)$	$= 9.1093837015\text{E-}31$	$e^2\,(c^2 \times 10^{-7})\,/\,(m_e^2\,G)$	$= 4.16560876666677\text{E+}42$
E_h	$E_h\,r_B\,/\,(r_e\,c^2)$	$= 9.1093837015\text{E-}31$	$E_h\,r_B\,/\,(m_e^2\,G)$	$= 4.16560876666677\text{E+}42$
G	$G\,(m_e^2\,S_G)\,/\,(r_e\,c^2)$	$= 9.1093837015\text{E-}31$	$G\,(m_e^2\,S_G)\,/\,(m_e^2\,G)$	$= 4.16560876666677\text{E+}42$
G_H	$G_H\,(m_e^2)\,/\,(r_e\,c^2)$	$= 9.1093837015\text{E-}31$	$G_H\,(m_e^2)\,/\,(m_e^2\,G)$	$= 4.16560876666677\text{E+}42$
S_G	$S_G\,(m_e^2\,G)\,/\,(r_e\,c^2)$	$= 9.1093837015\text{E-}31$	$S_G\,(m_e^2\,G)\,/\,(m_e^2\,G)$	$= 4.16560876666677\text{E+}42$
G_P	$G_P\,(m_e^2\,/\,S_G)\,/\,(r_e\,c^2)$	$= 9.1093837015\text{E-}31$	$G_P\,(m_e^2\,/\,S_G)\,/\,(m_e^2\,G)$	$= 4.16560876666677\text{E+}42$
F_N	$F_N\,(r_e^2\,S_G)\,/\,(r_e\,c^2)$	$= 9.1093837015\text{E-}31$	$F_N\,(r_e^2\,S_G)\,/\,(m_e^2\,G)$	$= 4.16560876666677\text{E+}42$
F_H	$F_H\,r_e^2\,/\,(r_e\,c^2)$	$= 9.1093837015\text{E-}31$	$F_H\,r_e^2\,/\,(m_e^2\,G)$	$= 4.16560876666677\text{E+}42$
F_P	$F_P\,(r_e^2\,/\,S_G)\,/\,(r_e\,c^2)$	$= 9.1093837015\text{E-}31$	$F_P\,(r_e^2\,/\,S_G)\,/\,(m_e^2\,G)$	$= 4.16560876666677\text{E+}42$

***

	The Speed of Light, c		The Bohr Velocity (Atomic Unit of Velocity), v_B	
m_e	$m_e (r_e c^2) / (m_e c r_e)$	= 299792458	$m_e (r_e c^2) / (m_e r_B v_B)$	= 2187691.26364101
r_e	$r_e (m_e c^2) / (m_e c r_e)$	= 299792458	$r_e (m_e c^2) / (m_e r_B v_B)$	= 2187691.26364101
t_e	$t_e (m_e c^3) / (m_e c r_e)$	= 299792458	$t_e (m_e c^3) / (m_e r_B v_B)$	= 2187691.26364101
r_B	$r_B (m_e v_B^2) / (m_e c r_e)$	= 299792458	$r_B (m_e v_B^2) / (m_e r_B v_B)$	= 2187691.26364101
t_B	$t_B (m_e v_B^3) / (m_e c r_e)$	= 299792458	$t_B (m_e v_B^3) / (m_e r_B v_B)$	= 2187691.26364101
v_B	$v_B (m_e r_B v_B) / (m_e c r_e)$	= 299792458	$v_B (m_e r_B v_B) / (m_e r_B v_B)$	= 2187691.26364101
$\hbar$	$\hbar v_B / (m_e c r_e)$	= 299792458	$\hbar v_B / (m_e r_B v_B)$	= 2187691.26364101
α	$\alpha (\hbar c) / (m_e c r_e)$	= 299792458	$\alpha (\hbar c) / (m_e r_B v_B)$	= 2187691.26364101
e^2	$e^2 (c^2 \times 10^{-7}) / (m_e c r_e)$	= 299792458	$e^2 (c^2 \times 10^{-7}) / (m_e r_B v_B)$	= 2187691.26364101
E_h	$E_h r_B / (m_e c r_e)$	= 299792458	$E_h r_B / (m_e r_B v_B)$	= 2187691.26364101
G	$G (m_e^2 S_G) / (m_e c r_e)$	= 299792458	$G (m_e^2 S_G) / (m_e r_B v_B)$	= 2187691.26364101
G_H	$G_H (m_e^2) / (m_e c r_e)$	= 299792458	$G_H (m_e^2) / (m_e r_B v_B)$	= 2187691.26364101
S_G	$S_G (m_e^2 G) / (m_e c r_e)$	= 299792458	$S_G (m_e^2 G) / (m_e r_B v_B)$	= 2187691.26364101
G_P	$G_P (m_e^2 / S_G) / (m_e c r_e)$	= 299792458	$G_P (m_e^2 / S_G) / (m_e r_B v_B)$	= 2187691.26364101
F_N	$F_N (r_e^2 S_G) / (m_e c r_e)$	= 299792458	$F_N (r_e^2 S_G) / (m_e r_B v_B)$	= 2187691.26364101
F_H	$F_H r_e^2 / (m_e c r_e)$	= 299792458	$F_H r_e^2 / (m_e r_B v_B)$	= 2187691.26364101
F_P	$F_P (r_e^2 / S_G) / (m_e c r_e)$	= 299792458	$F_P (r_e^2 / S_G) / (m_e r_B v_B)$	= 2187691.26364101

***

	The Classical Radius of the Electron, r_e		The Bohr Radius (Atomic Unit of Length), r_B	
m_e	$m_e (r_e c^2) / (m_e c^2)$	= 2.8179403262E-15	$m_e (r_e c^2) / (m_e v_B^2)$	= 2.41888432658569E-17
r_e	$r_e (m_e c^2) / (m_e c^2)$	= 2.8179403262E-15	$r_e (m_e c^2) / (m_e v_B^2)$	= 2.41888432658569E-17
t_e	$t_e (m_e c^3) / (m_e c^2)$	= 2.8179403262E-15	$t_e (m_e c^3) / (m_e v_B^2)$	= 2.41888432658569E-17
r_B	$r_B (m_e v_B^2) / (m_e c^2)$	= 2.8179403262E-15	$r_B (m_e v_B^2) / (m_e v_B^2)$	= 2.41888432658569E-17
t_B	$t_B (m_e v_B^3) / (m_e c^2)$	= 2.8179403262E-15	$t_B (m_e v_B^3) / (m_e v_B^2)$	= 2.41888432658569E-17
v_B	$v_B (m_e r_B v_B) / (m_e c^2)$	= 2.8179403262E-15	$v_B (m_e r_B v_B) / (m_e v_B^2)$	= 2.41888432658569E-17
$\hbar$	$\hbar v_B / (m_e c^2)$	= 2.8179403262E-15	$\hbar v_B / (m_e v_B^2)$	= 2.41888432658569E-17
α	$\alpha (\hbar c) / (m_e c^2)$	= 2.8179403262E-15	$\alpha (\hbar c) / (m_e v_B^2)$	= 2.41888432658569E-17
e^2	$e^2 (c^2 \times 10^{-7}) / (m_e c^2)$	= 2.8179403262E-15	$e^2 (c^2 \times 10^{-7})/(m_e v_B^2)$	= 2.41888432658569E-17
E_h	$E_h r_B / (m_e c^2)$	= 2.8179403262E-15	$E_h r_B / (m_e v_B^2)$	= 2.41888432658569E-17
G	$G (m_e^2 S_G) / (m_e c^2)$	= 2.8179403262E-15	$G (m_e^2 S_G) / (m_e v_B^2)$	= 2.4188843265869E-17
G_H	$G_H (m_e^2) / (m_e c^2)$	= 2.8179403262E-15	$G_H (m_e^2) / (m_e v_B^2)$	= 2.41888432658569E-17
S_G	$S_G (m_e^2 G) / (m_e c^2)$	= 2.8179403262E-15	$S_G (m_e^2 G) / (m_e v_B^2)$	= 2.41888432658569E-17
G_P	$G_P (m_e^2 / S_G) / (m_e c^2)$	= 2.8179403262E-15	$G_P (m_e^2/S_G) / (m_e v_B^2)$	= 2.41888432658569E-17
F_N	$F_N (r_e^2 S_G) / (m_e c^2)$	= 2.8179403262E-15	$F_N (r_e^2 S_G) / (m_e v_B^2)$	= 2.41888432658569E-17
F_H	$F_H r_e^2 / (m_e c^2)$	= 2.8179403262E-15	$F_H r_e^2 / (m_e v_B^2)$	= 2.41888432658569E-17
F_P	$F_P (r_e^2 / S_G) / (m_e c^2)$	= 2.8179403262E-15	$F_P (r_e^2 / S_G) / (m_e v_B^2)$	= 2.41888432658569E-17

***

Time Light takes to traverse the distance r_e, t_e			The Bohr Time (Atomic Unit of Time), t_B	
m_e	$m_e (r_e c^2) / (m_e c^3)$	= 9.39963715231288E-24	$m_e (r_e c^2) / (m_e v_B^3)$	= 2.41888432658569E-17
r_e	$r_e (m_e c^2) / (m_e c^3)$	= 9.39963715231288E-24	$r_e (m_e c^2) / (m_e v_B^3)$	= 2.41888432658569E-17
t_e	$t_e (m_e c^3) / (m_e c^3)$	= 9.39963715231288E-24	$t_e (m_e c^3) / (m_e v_B^3)$	= 2.41888432658569E-17
r_B	$r_B (m_e v_B^2) / (m_e c^3)$	= 9.39963715231288E-24	$r_B (m_e v_B^2) / (m_e v_B^3)$	= 2.41888432658569E-17
t_B	$t_B (m_e v_B^3) / (m_e c^3)$	= 9.39963715231288E-24	$t_B (m_e v_B^3) / (m_e v_B^3)$	= 2.41888432658569E-17
v_B	$v_B (m_e r_B v_B) / (m_e c^3)$	= 9.39963715231288E-24	$v_B (m_e r_B v_B)/(m_e v_B^3)$	= 2.41888432658569E-17
$\hbar$	$\hbar v_B / (m_e c^3)$	= 9.39963715231288E-24	$\hbar v_B / (m_e v_B^3)$	= 2.41888432658569E-17
α	$\alpha (\hbar c) / (m_e c^3)$	= 9.39963715231288E-24	$\alpha (\hbar c) / (m_e v_B^3)$	= 2.41888432658569E-17
E_h	$E_h r_B / (m_e c^3)$	= 9.39963715231288E-24	$E_h r_B / (m_e v_B^3)$	= 2.41888432658569E-17
G	$G (m_e^2 S_G) / (m_e c^3)$	= 9.39963715231288E-24	$G (m_e^2 S_G) / (m_e v_B^3)$	= 2.4188843265869E-17
G_H	$G_H (m_e^2) / (m_e c^3)$	= 9.39963715231288E-24	$G_H (m_e^2) / (m_e v_B^3)$	= 2.41888432658569E-17
S_G	$S_G (m_e^2 G) / (m_e c^3)$	= 9.39963715231288E-24	$S_G (m_e^2 G) / (m_e v_B^3)$	= 2.41888432658569E-17
G_P	$G_P (m_e^2 / S_G)/(m_e c^3)$	= 9.39963715231288E-24	$G_P (m_e^2/S_G)/(m_e v_B^3)$	= 2.41888432658569E-17
F_N	$F_N (r_e^2 S_G) / (m_e c^3)$	= 9.39963715231288E-24	$F_N (r_e^2 S_G) / (m_e v_B^3)$	= 2.41888432658569E-17
F_H	$F_H r_e^2 / (m_e c^3)$	= 9.39963715231288E-24	$F_H r_e^2 / (m_e v_B^3)$	= 2.41888432658569E-17
F_P	$F_P (r_e^2 / S_G) / (m_e c^3)$	= 9.39963715231288E-24	$F_P (r_e^2 / S_G)/(m_e v_B^3)$	= 2.41888432658569E-17

***

The Square of the Electron's Charge, e^2			The Hartree Energy, E_h	
m_e	$m_e (r_e c^2) / (c^2 \times 10^{-7})$	= 2.56696996792859E-38	$m_e (r_e c^2) / r_B$	= 4.3597447221905E-18
r_e	$r_e (m_e c^2) / (c^2 \times 10^{-7})$	= 2.56696996792859E-38	$r_e (m_e c^2) / r_B$	= 4.3597447221905E-18
t_e	$t_e (m_e c^3) / (c^2 \times 10^{-7})$	= 2.56696996792859E-38	$t_e (m_e c^3) / r_B$	= 4.3597447221905E-18
r_B	$r_B (m_e v_B^2) / (c^2 \times 10^{-7})$	= 2.56696996792859E-38	$r_B (m_e v_B^2) / r_B$	= 4.3597447221905E-18
t_B	$t_B (m_e v_B^3) / (c^2 \times 10^{-7})$	= 2.56696996792859E-38	$t_B (m_e v_B^3) / r_B$	= 4.3597447221905E-18
v_B	$v_B (m_e r_B v_B)/(c^2 \times 10^{-7})$	= 2.56696996792859E-38	$v_B (m_e r_B v_B) / r_B$	= 4.3597447221905E-18
$\hbar$	$\hbar v_B / (c^2 \times 10^{-7})$	= 2.56696996792859E-38	$\hbar v_B / r_B$	= 4.3597447221905E-18
α	$\alpha (\hbar c) / (c^2 \times 10^{-7})$	= 2.56696996792859E-38	$\alpha (\hbar c) / r_B$	= 4.3597447221905E-18
e^2	$e^2 (c^2 \times 10^{-7})/(c^2 \times 10^{-7})$	= 2.56696996792859E-38	$e^2 (c^2 \times 10^{-7}) / r_B$	= 4.3597447221905E-18
E_h	$E_h r_B / (c^2 \times 10^{-7})$	= 2.56696996792859E-38	$E_h r_B / r_B$	= 4.3597447221905E-18
G	$G (m_e^2 S_G) / (c^2 \times 10^{-7})$	= 2.56696996792859E-38	$G (m_e^2 S_G) / r_B$	= 4.3597447221905E-18
G_H	$G_H (m_e^2) / (c^2 \times 10^{-7})$	= 2.56696996792859E-38	$G_H (m_e^2) / r_B$	= 4.3597447221905E-18
S_G	$S_G (m_e^2 G) / (c^2 \times 10^{-7})$	= 2.56696996792859E-38	$S_G (m_e^2 G) / r_B$	= 4.3597447221905E-18
G_P	$G_P (m_e^2/S_G) / (c^2 \times 10^{-7})$	= 2.56696996792859E-38	$G_P (m_e^2 / S_G) / r_B$	= 4.3597447221905E-18
F_N	$F_N (r_e^2 S_G) / (c^2 \times 10^{-7})$	= 2.56696996792859E-38	$F_N (r_e^2 S_G) / r_B$	= 4.3597447221905E-18
F_H	$F_H r_e^2 / (c^2 \times 10^{-7})$	= 2.56696996792859E-38	$F_H r_e^2 / r_B$	= 4.3597447221905E-18
F_P	$F_P (r_e^2 / S_G) / (c^2 \times 10^{-7})$	= 2.56696996792859E-38	$F_P (r_e^2 / S_G) / r_B$	= 4.3597447221905E-18

***

Newton's Constant of Gravitation, G			The Newton Force Constant, F_N	
m_e	$m_e (r_e c^2) / (m_e^2 S_G)$	= 6.67430E-11	$m_e (r_e c^2) / (r_e^2 S_G)$	= 6.97461325379537E-42
r_e	$r_e (m_e c^2) / (m_e^2 S_G)$	= 6.67430E-11	$r_e (m_e c^2) / (r_e^2 S_G)$	= 6.97461325379537E-42
t_e	$t_e (m_e c^3) / (m_e^2 S_G)$	= 6.67430E-11	$t_e (m_e c^3) / (r_e^2 S_G)$	= 6.97461325379537E-42
r_B	$r_B (m_e v_B^2) / (m_e^2 S_G)$	= 6.67430E-11	$r_B (m_e v_B^2) / (r_e^2 S_G)$	= 6.97461325379537E-42
t_B	$t_B (m_e v_B^3) / (m_e^2 S_G)$	= 6.67430E-11	$t_B (m_e v_B^3) / (r_e^2 S_G)$	= 6.97461325379537E-42
v_B	$v_B (m_e r_B v_B) / (m_e^2 S_G)$	= 6.67430E-11	$v_B (m_e r_B v_B) / (r_e^2 S_G)$	= 6.97461325379537E-42
$\hbar$	$\hbar v_B / (m_e^2 S_G)$	= 6.67430E-11	$\hbar v_B / (r_e^2 S_G)$	= 6.97461325379537E-42
α	$\alpha (\hbar c) / (m_e^2 S_G)$	= 6.67430E-11	$\alpha (\hbar c) / (r_e^2 S_G)$	= 6.97461325379537E-42
e^2	$e^2 (c^2 \times 10^{-7}) / (m_e^2 S_G)$	= 6.67430E-11	$e^2 (c^2 \times 10^{-7}) / (r_e^2 S_G)$	= 6.97461325379537E-42
E_h	$E_h r_B / (m_e^2 S_G)$	= 6.67430E-11	$E_h r_B / (r_e^2 S_G)$	= 6.97461325379537E-42
G	$G (m_e^2 S_G) / (m_e^2 S_G)$	= 6.67430E-11	$G (m_e^2 S_G) / (r_e^2 S_G)$	= 6.97461325379537E-42
G_H	$G_H (m_e^2) / (m_e^2 S_G)$	= 6.67430E-11	$G_H (m_e^2) / (r_e^2 S_G)$	= 6.97461325379537E-42
S_G	$S_G (m_e^2 G) / (m_e^2 S_G)$	= 6.67430E-11	$S_G (m_e^2 G) / (r_e^2 S_G)$	= 6.97461325379537E-42
G_P	$G_P (m_e^2 / S_G) / (m_e^2 S_G)$	= 6.67430E-11	$G_P (m_e^2 S_G)/(r_e^2/S_G)$	= 6.97461325379537E-42
F_N	$F_N (r_e^2 S_G) / (m_e^2 S_G)$	= 6.67430E-11	$F_N (r_e^2 S_G) / (r_e^2 S_G)$	= 6.97461325379537E-42
F_H	$F_H r_e^2 / (m_e^2 S_G)$	= 6.67430E-11	$F_H r_e^2 / (r_e^2 S_G)$	= 6.97461325379537E-42
F_P	$F_P (r_e^2 / S_G) / (m_e^2 S_G)$	= 6.67430E-11	$F_P (r_e^2/S_G) / (r_e^2 S_G)$	= 6.97461325379537E-42

***

The Hatch Gauge Constant, G_H			The Hatch Force Constant, F_H	
m_e	$m_e (r_e c^2) / (m_e^2)$	= 2.7802522591364E+32	$m_e (r_e c^2) / (r_e^2)$	= 29.0535101141202
r_e	$r_e (m_e c^2) / (m_e^2)$	= 2.7802522591364E+32	$r_e (m_e c^2) / (r_e^2)$	= 29.0535101141202
t_e	$t_e (m_e c^3) / (m_e^2)$	= 2.7802522591364E+32	$t_e (m_e c^3) / (r_e^2)$	= 29.0535101141202
r_B	$r_B (m_e v_B^2) / (m_e^2)$	= 2.7802522591364E+32	$r_B (m_e v_B^2) / (r_e^2)$	= 29.0535101141202
t_B	$t_B (m_e v_B^3) / (m_e^2)$	= 2.7802522591364E+32	$t_B (m_e v_B^3) / (r_e^2)$	= 29.0535101141202
v_B	$v_B (m_e r_B v_B) / (m_e^2)$	= 2.7802522591364E+32	$v_B (m_e r_B v_B) / (r_e^2)$	= 29.0535101141202
$\hbar$	$\hbar v_B / (m_e^2)$	= 2.7802522591364E+32	$\hbar v_B / (r_e^2)$	= 29.0535101141202
α	$\alpha (\hbar c) / (m_e^2)$	= 2.7802522591364E+32	$\alpha (\hbar c) / (r_e^2)$	= 29.0535101141202
E_h	$E_h r_B / (m_e^2)$	= 2.7802522591364E+32	$E_h r_B / (r_e^2)$	= 29.0535101141202
G	$G (m_e^2 S_G) / (m_e^2)$	= 2.7802522591364E+32	$G (m_e^2 S_G) / (r_e^2)$	= 29.0535101141202
G_H	$G_H (m_e^2) / (m_e^2)$	= 2.7802522591364E+32	$G_H (m_e^2) / (r_e^2)$	= 29.0535101141202
S_G	$S_G (m_e^2 G) / (m_e^2)$	= 2.7802522591364E+32	$S_G (m_e^2 G) / (r_e^2)$	= 29.0535101141202
G_P	$G_P (m_e^2 / S_G) / (m_e^2)$	= 2.7802522591364E+32	$G_P (m_e^2 / S_G) / (r_e^2)$	= 29.0535101141202
F_N	$F_N (r_e^2 S_G) / (m_e^2)$	= 2.7802522591364E+32	$F_N (r_e^2 S_G) / (r_e^2)$	= 29.0535101141202
F_H	$F_H r_e^2 / (m_e^2)$	= 2.7802522591364E+32	$F_H r_e^2 / (r_e^2)$	= 29.0535101141202
F_P	$F_P (r_e^2 / S_G) / (m_e^2)$	= 2.7802522591364E+32	$F_P (r_e^2 / S_G) / (r_e^2)$	= 29.0535101141202

***

	The Planck Gauge Constant, G_P		The Planck Force Constant, F_P	
m_e	$m_e (r_e c^2) / (m_e^2/ S_G)$	1.15814431842037E+75	$m_e (r_e c^2) / (r_e^2 / S_G)$	1.21025556433821E+44
r_e	$r_e (m_e c^2) / (m_e^2/ S_G)$	1.15814431842037E+75	$r_e (m_e c^2) / (r_e^2 / S_G)$	1.21025556433821E+44
t_e	$t_e (m_e c^3) / (m_e^2/ S_G)$	1.15814431842037E+75	$t_e (m_e c^3) / (r_e^2 / S_G)$	1.21025556433821E+44
r_B	$r_B (m_e v_B^2) / (m_e^2/ S_G)$	1.15814431842037E+75	$r_B (m_e v_B^2)/ (r_e^2 / S_G)$	1.21025556433821E+44
t_B	$t_B (m_e v_B^3) / (m_e^2/ S_G)$	1.15814431842037E+75	$t_B (m_e v_B^3)/ (r_e^2 / S_G)$	1.21025556433821E+44
v_B	$v_B (m_e r_B v_B)/(m_e^2/ S_G)$	1.15814431842037E+75	$v_B (m_e r_B v_B)/(r_e^2/S_G)$	1.21025556433821E+44
$\hbar$	$\hbar\, v_B / (m_e^2/ S_G)$	1.15814431842037E+75	$\hbar\, v_B / (r_e^2 / S_G)$	1.21025556433821E+44
α	$\alpha (\hbar c) / (m_e^2/ S_G)$	1.15814431842037E+75	$\alpha (\hbar c) / (r_e^2 / S_G)$	1.21025556433821E+44
E_h	$E_h\, r_B / (m_e^2/ S_G)$	1.15814431842037E+75	$E_h\, r_B / (r_e^2 / S_G)$	1.21025556433821E+44
G	$G (m_e^2 S_G) / (m_e^2/ S_G)$	1.15814431842037E+75	$G (m_e^2 S_G)/(r_e^2 / S_G)$	1.21025556433821E+44
G_H	$G_H (m_e^2) / (m_e^2/ S_G)$	1.15814431842037E+75	$G_H (m_e^2) / (r_e^2 / S_G)$	1.21025556433821E+44
S_G	$S_G (m_e^2 G) / (m_e^2/ S_G)$	1.15814431842037E+75	$S_G (m_e^2 G)/(r_e^2 / S_G)$	1.21025556433821E+44
G_P	$G_P (m_e^2/S_G)/ (m_e^2/ S_G)$	1.15814431842037E+75	$G_P (m_e^2/S_G)/(r_e^2/S_G)$	1.21025556433821E+44
F_N	$F_N (r_e^2 S_G) / (m_e^2/ S_G)$	1.15814431842037E+75	$F_N (r_e^2 S_G) / (r_e^2 / S_G)$	1.21025556433821E+44
F_H	$F_H\, r_e^2 / (m_e^2/ S_G)$	1.15814431842037E+75	$F_H\, r_e^2 / (r_e^2 / S_G)$	1.21025556433821E+44
F_P	$F_P (r_e^2 / S_G)/ (m_e^2/ S_G)$	1.15814431842037E+75	$F_P (r_e^2 / S_G)/(r_e^2/S_G)$	1.21025556433821E+44

$$***$$

A closer look at the current situation is warranted. The G_H derivations of the ground-state Bohr model are essential to the exact-numbers, MaxD-Ratio Base. Yet the facts relating to the roles of Maxwell's mass dimensions $[L^3\, T^{-2}]$, and his dimensions $[L^3\, T^{-2}\, M^{-1}]$, of Newton's G, are essential to any sensible explanation of what we have demonstrated. The fact is that dividing by G_H simply replaces Maxwell's dimensions of mass, $[L^3\, T^{-2}]$, our (r_e^3 / t_e^2) or $(c^2\, r_e)$, with the actual rest mass of the electron.

This fact alone, answers objections such as Leon Lederman's as to the validity of the Bohr model.

On page 163 of his "The God Particle" Lederman wrote: " . . . the now greatly aggrandized Bohr theory, . . . could account for a very impressive amount of precise and brilliantly achieved experimental data. There was only one problem. It was wrong."

Without awareness of what the G_H derivations of the Bohr model tell us about the essential role of the Maxwell dimensions, it would be difficult for almost any modern physicist to accept the implications.

Yet, the MaxD Physics is an Exact-Numbers physics that tells us that the numbers are real. Dividing by G_H does indeed accomplish the replacement of (r_e^3 / t_e^2) or $(c^2\, r_e)$, with the rest-mass of the electron.

Just so, the Hatch gauge constant, G_H, has freed the Bohr model from the common perception that the model was somehow "wrong" even though its ability to account for a very impressive amount of precise experimental data could never be challenged.

The classical radius of the electron and SI's recommended derivation provides validation of our numbers. A few identities provide a background.
- $(m_e\, r_e) = (e^2 \times 10^{-7})$ This reality is found in the SI data base, giving us a running start.
- $r_e = (e^2 \times 10^{-7}) / m_e$ $m_e = (e^2 \times 10^{-7}) / r_e$ $e^2 = ((m_e\, r_e) \times 10^7)$

The SI specified derivation of r_e is detailed with respect to relationships with other constants.

- $r_e = (1 / (4\pi \, \varepsilon_0)) \, (e^2 / (m_e \, c^2))$ $= (e^2 / (4\pi \, \varepsilon_0)) / (m_e \, c^2)$ $\quad r_e = e^2 \, (c^2 \times 10^{-7}) / (m_e \, c^2)$
- Dividing by r_e: $\quad 1 = e^2 \, (c^2 \times 10^{-7}) / (m_e \, c^2 \, r_e)$
- We have two definitions of our Electro-Gravito constant: $\quad E_G = e^2 \, (c^2 \times 10^{-7}) = (m_e \, c^2 \, r_e)$.

Countless additional verifications are easy to find.

I now look to the past. This may help explain why I have so much confidence in the New Physics that is now available. I first present the history of additions to my toolbox over the last twenty years. The table presents them as a list of electron-scale constants I have added to my MaxD-Ratio toolbox.

Exact values of constants I have added to the MaxD toolbox.			
t_e	$= (r_e / c)$	= 9.39963715231288E-24	time for light to traverse r_e.
S_G	$= (G_H / G)$	= 4.16560876666677E+42	The Gauge Symmetry Constant
G	$= (G_H / S_G)$	= 6.67430E-11	Newton's Constant of Gravitation
G_H	$= (r_e^3 / t_e^2) / m_e$	= 2.7802522591364E+32	The Hatch Gauge Constant
G_P	$= (G_H \, S_G)$	= 1.15814431842037E+75	The Planck Gauge Constant
F_N	$= (F_H / S_G)$	= 6.97461325379537E-42	The Newton Force Constant
F_H	$= (m_e^2 / r_e^2)$	= 29.0535101141202	The Hatch Force Constant
F_P	$= (F_H \, S_G)$	= 1.21025556433821E+44	The Planck Force Constant

By 2001 I had added my time, t_e, my dimensions of G, G_H, and S_G, my dimensionless gauge symmetry constant. Together they allowed me to define G in terms of G_H. In 2002 I used G_H and S_G to present a simplified version of the Planck units of length, time, and mass.

Essential Redefinitions In Base Units $\hbar$, G, and c, are:
$\hbar = (m_e \, c \, r_e / \alpha)$
$G = (r_e \, c^2 / m_e) / S_G$
$c = (r_e / t_e)$

Simplifications of the Planck units	Planck	Hatch
l_P = Length	$= (\hbar \, G / c^3)^{(1/2)}$	$= r_e / (S_G \, \alpha)^{(1/2)}$
t_P = Time	$= (\hbar \, G / c^5)^{(1/2)}$	$= t_e / (S_G \, \alpha)^{(1/2)}$
m_P = Mass	$= (\hbar \, c / G)^{(1/2)}$	$= m_e \, (S_G / \alpha)^{(1/2)}$

The squares of the respective identical values, Planck vs Hatch, also provided significant results.

The Planck length, l_P		The Planck time, t_P		The Planck mass, m_P	
l_P	= 1.61625502442062E-35	t_P	= 5.3912464483033E-44	m_P	= 2.17643434271374E-08

A Length Ratio	A Time Ratio	A Mass Ratio
$\alpha = (r_e^2 / l_P^2) / S_G$	$\alpha = (t_e^2 / t_P^2) / S_G$	$\alpha = S_G \, (m_e^2 / m_P^2)$
α = 0.00729735256929315	α = 0.00729735256929315	α = 0.00729735256929315

In any context all fine structure constant derivations can be reduced to $\alpha = (v_B / c)$.

Next, simply because I could, I decided to change the conventional Planck unit derivations so that I could use the same form to derive pure electron-based units. It was remarkably easy to get what I wanted.

By simply replacing $\hbar$ with $(\hbar\,\alpha)$ and G with G_H I had:

Length	$(\hbar\,\alpha\,G_H\,/\,c^3)^{(1/2)}$	$= r_e$	$= 2.8179403262E\text{-}15$
Time	$(\hbar\,\alpha\,G_H\,/\,c^5)^{(1/2)}$	$= t_e$	$= 9.39963715231288E\text{-}24$
Mass	$(\hbar\,\alpha\,c\,/\,G_H)^{(1/2)}$	$= m_e$	$= 9.1093837015E\text{-}31$
Force	$(c^4\,/\,G_H)$	$= F_H$	$= 29.0535101141202$

These were the pictures in my 2002 presentation offered to all interested parties.

The additional Gauge and Force constants below, (circa 2015), seem to provide a limited unification of gravitational and electromagnetic units. I spent some time examining it, but other priorities prevailed.

Playing with Gauge and Force constants.

The three "Gauge Constants" and the associated "Force Constants" provide a surprising picture.					
Newton Gauge, $G = (G_H\,/\,S_G)$		Hatch Gauge, $G_H = (r_e^3\,/\,t_e^2)\,/\,m_e$		Planck Gauge, $G_P = (G_H\,S_G)$	
F_N	$= (F_H\,/\,S_G)$	$=$ The Newton Force　　The Planck Force $=$		$= c^4\,/\,G)$	$= F_P$
F_H	$= G_H\,(m_e^2\,/\,r_e^2)$	$=$ The Hatch Force $=$		$= c^4\,/\,G_H)$	$= F_H$
F_P	$= (F_H\,S_G)$	$=$ The Planck Force　　The Newton Force $=$		$= c^4\,/\,G_P$	$= F_N$
These Gauge and Force constants tell a unique new story, begging for a discussion of implications.					

The MaxD Ratio. I present a closer look at the origin and use of the ratio. In 2002 I had already become convinced of the fundamental importance of m_e, c, and r_e. I needed a much better understanding of what I was playing with.

I first explored using the four constants that were clearly related to the $(m_e\,c^2\,r_e)$. Obviously:

- m_e 　　$= (m_e\,c^2\,r_e)\,/\,(c^2\,r_e)$ 　　$= m_e$ 　　giving a constant, 　　$m_e\,(c^2\,r_e)$.
- c 　　$= (m_e\,c^2\,r_e)\,/\,(m_e\,c\,r_e)$ 　　$= c$ 　　giving a constant, 　　$c\,(m_e\,c\,r_e)$.
- r_e 　　$= (m_e\,c^2\,r_e)\,/\,(m_e\,c^2)$ 　　$= r_e$ 　　giving a constant, 　　$r_e\,(m_e\,c^2)$.
- t_e 　　$= (m_e\,c^2\,r_e)\,/\,(m_e\,c^3)$ 　　$= t_e$ 　　giving a constant, 　　$t_e\,(m_e\,c^3)$.

The constants on the right were all $(m_e\,c^2\,r_e)$. Obviously, I did not need a calculator to assure me that these results were correct. Other constants could not be so easy. I needed to fully understand the apparently easy process. I used Planck's reduced constant as a test of possible further use of the process:

- $(m_e\,c^2\,r_e)\,/\,\hbar = z$ 　　$(m_e\,c^2\,r_e) = (\hbar\,z)$ 　　$\hbar = ((m_e\,c^2\,r_e)\,/\,z)$
- I finally realized that $(z = (c\,\alpha))$ and is the Bohr velocity, $v_{B,}$.
- $\hbar = ((m_e\,c^2\,r_e)\,/\,v_B)$ 　　$= ((m_e\,c^2)\times(r_e\,/\,v_B))$ 　　$=$ (Energy x Time)
- $(m_e\,c^2\,r_e)$ 　　$= ((m_e\,c^2)\times(r_e\,/\,v_B)\times v_B$ 　　$=$ (Energy x Time x velocity)

It was not easy, and why should "z" be a generally recognized factor or factors of $\hbar$? I did not know, But it is easy to see how encountering $E_G = (m_e\,c^2\,r_e)$ some years later turned me to my old MaxD Ratio. It seemed very much like magic given the magnitude of the number of implied new relationships. Like everyone else, I had not realized that the cgs e^2 was precisely my $(m_e\,c^2\,r_e)$.

Finally, over the past few years I have published a small book and sent more than one-hundred copies of my work to anyone interested. The above comparison of values makes it appear that key parts of my work have filtered into NIST and CODATA awareness. Their rather sudden interest in precise non-experimental values is very new. Continued inconsistencies may have triggered a search for a better, not totally experimentally constrained, numbers base. Exact numbers are now available.

An Exact-Number Base, can not tolerate what I call "pi-pollution". An observant reader may have noticed that no factor of pi is found in our exact-number base. The only place one will find the term mentioned to this point is in the: $e^2 / (4pi \, \varepsilon_0)$ $= e^2 (c^2 \times 10^{-7}) = (m_e \, c^2 \, r_e)$, definition of E_G, and it is obvious there that our base is pi-pollution free. Mass, length, time, velocity, and charge, provide the MaxD base.

- **Mass:** m_e, the mass of the electron is $(m_e \, c^2 \, r_e) / (c^2 \, r_e)$
- **Length:** r_e, the classical radius of the electron is $(m_e \, c^2 \, r_e) / (m_e \, c^2)$ $\qquad = (c \; t_e)$
- **Time:** t_e, is $(m_e \, c^2 \, r_e) / (m_e \, c^3)$. $\qquad = (r_e / c)$
- **Velocity:** c, is the speed of light $(m_e \, c^2 \, r_e) / (m_e \, c \, r_e)$ $\qquad = (r_e / t_e)$
- **Charge.** e, is the charge of the electron $(e^2 (c^2 \times 10^{-7})^{(1/2)}$

One of the 2018 CODATA mysteries is the claim that Planck's constant is exact. That is clearly not true. With pi as a factor a number is never truly exact. Observe:

- Planck's constant is 2pi times the reduced constant, h-bar.
- CODATA gives Planck's "exact" constant to only nine digits.
- Yet my software gives 2pi as 2pi = 6.28318530717959.
- At nine digits 2pi = 6.28318530
- Again, in my software, (2pi / 6.28318530) = 1.00000000114267. This is far from exact.

Of course, none of this precludes the freedom to use pi wherever one wants to or wherever it is needed. Just don't pretend it can be present in an exact-numbers base. Don't ever, ever, ever, call it exact. It is not. Exact values are available for mass, length, time, velocity, and charge. That is enough—now use pi wherever you wish. No problem? Well . . . You may see inconsistencies if others work at a different precision. But . . .

Welcome to The New Physics

We have introduced significant aspects of a new, exact-numbers, theoretical physics base.

1. Exact values. MaxD physics has given both consistent and exact numbers. The consistency of exact values has never been matched.

2. **Previously unsuspected relationships.** The magnitude of the number of new relationships will result in countless new works examining them.

3. **Credibility of the Bohr model is explained.** The implications of the new visibility of Maxwell's dimensions of mass and of G is truly overwhelming.

4. **A New understanding of the fine structure constant.** The Bohr model as well as other evidence such as the length, time, and mass, ratios, demonstrated by the square of the Planck units, forever tell us that in every context the fine structure constant can easily be reduced to $\alpha = (v_B / c)$.

5. A new exact numbers base. This may require re-education on a massive scale.

Any comments with respect to this work are welcomed at: electrophy.ed@gmail.com